Creating Your Earth-Friendly Early Childhood Program

Other Redleaf Press Books by Patty Born Selly

Connecting Animals and Children in Early Childhood

Early Childhood Activities for a Greener Earth

Teaching STEM Outdoors: Activities for Young Children

Creating Your Earth-Friendly Early Childhood Program

Patty Born Selly

Published by Redleaf Press
10 Yorkton Court
St. Paul, MN 55117
www.redleafpress.org

First edition 2022
Senior editor: Melissa York
Cover design by Renee Hammes
Cover photos by pingpao / Nickyt87 / Oksana Kuzmina / stock.adobe.com
Typeset in Signo and Avenir by Douglas Schmitz
Printed in the United States of America
29 28 27 26 25 24 23 22 1 2 3 4 5 6 7 8

Library of Congress Cataloging-in-Publication Data

Names: Selly, Patty Born, author.
Title: Creating your earth-friendly early childhood program / by Patty Born Selly.
Description: First edition. | St. Paul, MN : Redleaf Press, 2022. | Includes bibliographical references. | Summary: "Creating your Earth-Friendly Early Chlidhood Program, Redleaf Quick Guide offers an approachable, efficient entry point for ECE educators who wish to instill ecofriendly values and practices in their programs. The guide will help educators evaluate their current environment and practices, get families and colleagues involved, and make both immediate and long-term changes to make their program "greener."— Provided by publisher.
Identifiers: LCCN 2022007442 (print) | LCCN 2022007443 (ebook) | ISBN 9781605547695 (paperback) | ISBN 9781605547701 (ebook)
Subjects: LCSH: Early childhood education--United States--Administration. | Early childhood education--Environmental aspects--United States. | Environmental responsibility--United States. | Sustainable living--United States. | Children and the environment--United States. | Environmental education--United States.
Classification: LCC LB2822.6.S45 2022 (print) | LCC LB2822.6 (ebook) | DDC 333.7071--dc23/eng/20220314
LC record available at https://lccn.loc.gov/2022007442
LC ebook record available at https://lccn.loc.gov/2022007443

Printed on acid-free paper

CONTENTS

CHAPTER 1
WHAT DOES IT MEAN TO GO GREEN?

Since approximately the late 1990s, the number of practitioners and administrators in early childhood education looking to "go green" has increased. Parents and caregivers across the United States and beyond are increasingly interested in programs that emphasize Earth-friendly practices and curricula that involve behaviors such as recycling, using environmentally friendly materials, creating schoolyard gardens, and spending more time outdoors. Programs around the United States—urban and rural, corporate, locally owned, or home-based—are moving to adopt so-called "green" practices in response to this growing demand. Early care and education programs are striving to be healthy settings where parents and caregivers can be confident, knowing their children are in good hands. Whether your program is located in a nature-rich area or has very little access to green space, you can use the outdoor environment and other resources around you to support your eco-friendly goals, whatever they may be.

How This Book Works

If you're reading this Quick Guide, there is a strong possibility that you're interested in reducing the environmental impact of your program and adopting more eco-friendly practices. You're not alone. Thousands of programs across the United States and beyond are engaged in this work in a variety of ways.

This book offers advice to help practitioners, administrators, and other caregivers adopt more eco-friendly practices and policies throughout their programs as a response to the demands of busy, concerned families. The purpose of this book is to meet you wherever you are on your journey to becoming a more eco-friendly program and

- provide examples of what an eco-friendly program looks like, with lots of big and small ways to shift your own program along the continuum;

You can apply the term *Earth-friendly* or *eco-friendly* to classroom practices, to building and site operations, and anything in between. As you proceed through this Quick Guide, it will be helpful to reflect on what the label *Earth friendly* means to you, your colleagues and staff, and the parents in your community. Another common term is *sustainability*. Sustainability is commonly defined as the ability to meet our needs without compromising the needs of future generations—others explain it more simply as living within our means. You might also think about sustainability as it relates to the health of children, adults, families, neighborhoods, communities, the natural world, and the planet. Sustainability can apply to behaviors, practices, and policies as well. Do your practices do more good than harm? Do they support families and promote wellness? Do they honor the integrity and well-being of families and communities? What about the natural world?

Why Go Green?

The purpose of this guide is to help you identify ways you can be successful in whatever choices you make toward your goal to become more eco-friendly. Just as there is a wide assortment of ways that individuals can do good things for the environment, many ways exist for programs to be more eco-friendly: through operational changes, curricular adaptations, and more. This book will help you identify strategies and practices that will work in your setting regardless of your role, position, and background knowledge. You may have a solid understanding of environmental issues, or you may be new to all of this, but rest assured that you are not alone. Countless teachers, paraprofessionals, and program administrators throughout the country are learning about environmental issues and striving to make their programs more Earth friendly. They chose to do so with confidence that these practices ultimately support children's health, wellness, and social relationships to ensure a brighter future for all.

Most people—even young children to some extent—are aware of the many environmental issues that confront us: air and water pollution, climate change, and loss of biodiversity across the planet. Many of these issues have direct effects on human health and well-being; rising asthma rates, high concentrations of pollutants or toxic chemicals in food and drinking water, and even wildly fluctuating weather patterns have serious impacts. Parents are understandably concerned for their children's future. Knowing that many of these environmental problems can lead to health issues, parents want to know that wherever their children spend their time, they are doing it in places that have clean water, healthy air, and indoor environments that are free from germs and other toxins.

A common element in many programs' shift toward environmentally oriented lifestyles is a recognition of health and wellness. Parents want to support their children in growing, learning, and being as healthy as they can be. Offering healthy food choices such as more fruits and vegetables at snacktime, increasing children's outdoor time and physical activity, and teaching coping skills such as meditation and mindfulness are some of the ways early care and education programs seek to build healthy lifestyles for children. Especially in response to COVID-19, many programs have made significant operational changes to improve cleaning and sanitizing practices, including being more mindful about handwashing throughout the day.

Moreover, adults and children alike generally want to do positive things for the planet, a result of hearing media reports, reading children's books, and reflecting on their own lived experience. In response to the growing awareness surrounding environmental issues, many parents want their children to learn about nature and animals so they develop feelings of stewardship and care for the environment. Parents and other adults feel concerned about the future and want children to grow up in a world where there is access to nature, clean water, and fresh air.

A significant focus on nature, the environment, or time outdoors is very generally referred to as *environmental education*. One purpose of environmental education is to "chart an appropriate and positive process . . . [to] start young children on their journey toward becoming environmentally responsive youth and adults" (NAAEE 2010, 3). Plenty of research supports the use of the environment as a setting for learning: time in nature has been shown to improve academic outcomes, foster resilience, and improve social-emotional relationships, among a host of other benefits.

Program administrators and funders will also be pleased to know that many so-called green practices carry significant financial benefits too. Many energy-saving operational practices can be steps toward sustainability. Improving heating and cooling efficiency, conserving water, updating waste management practices, and using consumable resources such as cleaning products and packaging more efficiently can lead to real savings if done thoughtfully.

Earth-Friendly Choices and Equity

All children and families deserve healthy, safe, Earth-friendly places to learn, grow, and be together. All children deserve the opportunity to play outdoors and celebrate the natural world. It is critically important that we who work with young children—whatever the setting—respond to this need and honor children and families by doing as much as possible to support thriving, Earth-friendly, justice-oriented communities and classrooms. The United Nations Children's Fund explains how climate change affects children specifically:

> The last 10 years were the hottest on record and the number of climate-related disasters has tripled in the last 30 years. These disasters have a disproportionate impact

CHAPTER 2
THE INDOOR ENVIRONMENT

Mr. Lopez believes that caring for children is his life's calling. He not only wants to support them in their learning and development but also wants them to be as healthy as possible. He and his staff model positive behaviors like frequent handwashing, making healthy food choices, and more. Mr. Lopez knows there is more to a healthy environment than just what's visible. He has thought a lot about the indoor environment, since children spend so much time indoors every day. Mr. Lopez researched products, including soaps, cleaning products, and wipes, to find those that have the fewest harmful chemicals. He and his staff make their own air fresheners and keep the windows open as much as possible to improve airflow. Children and adults alike remove their street shoes and put on indoor shoes as they enter the classroom, so as not to track in any harmful chemicals they may have on their outdoor shoes.

You may not wish to adopt every practice that Mr. Lopez has incorporated into his program, but this vignette provides some examples of the many ways in which program administrators and teachers can support children's health through the everyday choices they make regarding hygiene and cleaning in early childhood programs.

Environmental Health

Caring for our own health and that of others is central to early childhood caregiving. Practitioners work to model healthy behavior through frequent handwashing, healthy eating, physical activity, and consistent, rigorous cleaning and sanitization processes. Special precautions such as using sunscreen and wearing proper outdoor clothing, following cleaning practices specific to outdoor toys, and enclosing play areas in barriers or fences ensure children's health and safety during outdoor play. Since so many early childhood programs are indoors for the majority of the time, this Quick Guide focuses on some of the factors that contribute to the indoor health of the program environment and influence the quality of our own health and that of the children in our care.

An awareness of environmental health can have a pretty big influence on one's approach to making eco-friendly choices. As described in my book *Early Childhood Activities for a Greener Earth*, the phrase *environmental health* encompasses elements that influence the overall quality of an indoor environment, specifically as it relates to human health. These include chemicals, air quality, water quality, food quality, and even the supplies and materials you use every day. In early childhood settings, some specific factors that affect environmental health include cleaning products and personal care products such as soaps and hand sanitizers.

As practitioners model pro-environmental behavior and teach children about the importance of caring for their health through proper handwashing and healthy eating, they also teach children about caring for the physical spaces they inhabit as well as the environment as a whole. Those who work in early childhood know that modeling has an enormous effect on children's learning and engagement: simply put, when children see adults demonstrating behaviors, attitudes, or habits, they are much more inclined to do so themselves. Prominent researchers have determined that when children have positive, nurturing experiences with the natural world, they are much more likely to grow into youth and adults who are invested in caring for the natural world and acting on behalf of the creatures that inhabit it (Chawla and Derr 2012). In this way, being thoughtful about caring for one's body translates to being thoughtful about one's physical environment, which leads to behaviors that demonstrate care and responsibility for the environment and planet too. Of course, this doesn't just happen. Teachers need to talk openly about the choices they are making so that simple acts such as cleaning become demonstrations of care for the classroom and the people, pets, and plants within it. When it's time to prepare snack, educators can explain how the food choices are free of chemicals and are locally grown so they have minimal impact on the Earth (about foods for which this is true, of course) and talk about how they help children grow strong and be healthy. To help children understand the context of where their food comes from, teachers may also even express gratitude to the earth and to the farmers and workers who made the food accessible.

You want to have the cleanest, healthiest environment possible for the children in your care. You want to reduce the presence of bacteria, viruses, germs, and insects to decrease children's potential exposure to illness and other risks. Certain cleaning routines and practices are required by law in some states, and in response to the COVID-19 pandemic, many state licensing requirements have tightened up regulations about cleaning and sanitizing practices. As well, certain chemicals are more effective than others at eliminating viruses. For more information, visit the Centers for Disease Control or the Environmental Protection Agency's page on COVID-19 for specific details related to the coronaviruses (the family of viruses that causes COVID-19, as well as common colds and the flu). This may make it impossible to remove some chemicals from your environment. However, if you want to make shifts to be more environmentally benign, it pays to start by being aware of what is in use at your program.

This chapter provides an overview of some of the common environmental health concerns in early childhood settings. It starts with a brief description of the health hazards associated with common substances such as cleaning chemicals and plastics. Next, it identifies eco-friendly approaches to consider for maintaining your indoor environment, offering practical strategies for practices that are in keeping with your tried-and-true safety and cleaning protocols while at the same time helping you and your colleagues shift to a greener approach.

Environmental Toxins and Effects

Chemicals are present everywhere. They make up our bodies as well as the products and substances we consume. Some chemicals are benign or mostly harmless, although they can sometimes be problematic. Unfortunately, it's often difficult to discern the exact chemical makeup of many of the cleaning products we use daily. Laws in the United States allow corporations to keep many ingredient lists secret, citing the right of companies to hold proprietary information. Even when chemicals pass extensive testing and are shown to be nontoxic in adult humans, sometimes they affect children more, building up in the fatty tissues of the body and adding to what's known as a *body burden* (Quinn and Wania 2012). That term defines the cumulative effect of chemicals in the body.

Young children go through periods of brain and body development during which they are more susceptible to the negative effects of chemicals. In other words, something that may not have a measurable negative effect on a twelve-year-old could have a significant negative effect on a three-year-old, due to the massive developmental changes happening throughout the younger child's body throughout early childhood. These time periods in a child's development are referred to as *critical windows*, a period of time when children's vulnerability to the effects of harmful chemicals is high (Lanphear 2015). Critical windows fall during different developmental phases in different children, but generally speaking, the ages from infancy through about age seven are extremely vulnerable times. In addition, factors such as the child's home environment and previous chemical exposures, genetic makeup, diet, and racial and socioeconomic status all affect a child's susceptibility to chemical exposure.

Since the advent of COVID-19, we've all become even more vigilant in ensuring that surfaces are clean, disinfected, and free from bacteria and viruses. As many early childhood care and education settings have increased or intensified their cleaning and disinfecting procedures, it's of paramount importance to be aware of the potential harmful effects of the chemicals we use to protect us. For example, household bleach is a commonly used substance, frequently diluted with water to create a powerful disinfectant. But what makes bleach such a powerful cleaning agent can burn skin and irritate eyes, nasal passages, lungs, and other mucous membranes, and in some cases disrupt certain kinds of hormones that are critical for development. Some evidence suggests that triclosan, a chemical that's commonly found in products such as hand sanitizer, mouthwash, and toothpaste, may lead to reproductive harm later in life if ingested (Weatherly and Gosse 2017).

Homemade Cleaning Products

Many household cleaners are easy to swap out with fresh-smelling, homemade, effective alternatives that do the job just as well but are much safer and much better for the environment. Store them in the fridge to preserve freshness. The ingredients here aren't listed with the EPA as disinfectants, so check with your licensing agent to see where you can use them appropriately at your site.

1. All-purpose cleaner: Mix equal parts white vinegar and water in a spray bottle. For extra cleaning power, add a little mild dish liquid (about 1 teaspoon per 32 oz. spray bottle) You can add scents like fresh herbs, lemon or lime peel, or essential oils if desired. Rinse the surface after using.
2. Scouring cleaner: Mix 1/3 cup baking soda with ¼ cup water and a tablespoon of lemon juice to make a thick paste. This scrub works well as a replacement for bleach and more abrasive cleaners.
3. Deodorizer sachets: Fill the feet of thin, clean socks with dried herbs such as lavender and rosemary and add baking soda. Tie tightly with twine and hang in coat closets or storage containers.

Children's Exposure

Children can be exposed to harmful chemicals in many ways, but three of the most common pathways are ingestion, absorption, and inhalation. While children in child care certainly don't have access to eat or drink common cleaning chemicals, they still ingest them when they consume the lingering residue that remains on surfaces after they've been cleaned. For example, if you spray a disinfectant on a table or chair and neglect to rinse it, children's hands and arms pick up minute amounts of chemicals on their skin. And we all know how much children love to put their hands into their mouths! After resting their hands on a "freshly cleaned" table while coloring a picture, a small child who later sucks their thumb is consuming any cleaning residue that happens to linger on their skin. Chemicals in floor cleaners and carpet cleaners and contaminants that come in on peoples' shoes (including pesticides or heavy metals that build up in soils) are also common culprits. Given the amount of time young children spend on the ground, it is a smart practice to avoid toxic cleaning chemicals on the floor and ask people to remove their shoes indoors.

Takeaway Tip

It is important to rinse all surfaces that have been cleaned with chemicals. Even sprays that proclaim "No rinsing required!" should be rinsed after application to avoid children's accidental ingestion of minute quantities through residual remains.

Another way that children take in chemicals is by absorbing them through their skin. The skin's job is to be a protective barrier, and yet it is also porous. Children's skin is delicate and sensitive, making them more susceptible to absorption than adults. Soaps, lotions, sunscreens, mosquito repellant: all of these common products can contain questionable chemicals.

Children also consume toxic chemicals as they inhale tiny particles or aerosolized droplets in the air. Anytime you can smell scents, fumes, or liquids in the air, you are taking them into your bloodstream simply by inhaling them. Since children's bodies are smaller, the effects of the harmful chemicals may be more profound. For this reason, it's best to avoid perfuming the air with unnecessary scented cleaners, air fresheners, and other products that linger in the air. There are many natural ways of freshening the air, such as running fans or air purifiers, opening windows, and making nontoxic air fresheners.

For ease of use, many cleaning supplies such as disinfectants and personal care products such as sunscreens come in aerosol containers. However, this method of application can be hazardous because it offers less control over the product. Aerosol containers contain propellants—gases stored under pressure that help with application. These propellants do just what their name suggests; they propel the chemicals out of the container, onto the surface, into the surrounding air, and into delicate lungs. This makes them difficult to contain, as the tiny droplets spread far and wide. If you must use aerosols, I recommend spraying the product onto a cleaning rag or cloth first and then wiping the surface. While this won't completely contain the particles, it will give you a measure of control.

Common Chemical Exposure

When and where are children exposed to toxic chemicals?

- Everyday chemicals that are often used in early care and education settings include drain cleaners, abrasive cleaners, bleach, window sprays, carpet cleaners, laundry detergents, and insecticides.
- Substances used directly on the skin by children and staff include soaps, hand sanitizer, sunscreen, mosquito repellent, and wet wipes.
- Children are also exposed to harmful chemicals in common products such as plastic bottles and toys, shampoos and soaps, plush toys, and even some medicines.
- Fabrics, furniture, curtains, carpets, pads, mats, and paints can all emit fumes in a process known as off-gassing. Fumes from these materials can linger in the air for days after they are installed.

The chemical compositions of these materials vary widely and often are not disclosed to consumers. It is not within the scope of this book to offer an exhaustive list, but you can get started with the resources listed in the More Information section at the end of this chapter.

Reading Labels

With all of this information, you may be wondering how to sort out and digest it to keep the children in your care safe and healthy. While it's always a good idea to do your research on specific products and manufacturers as suggested here, a good way to start is simply by making a habit of reading product labels. At times it can seem overwhelming, especially if the names of ingredients are words you don't recognize. Without going back to school for a chemistry degree, one way to assess the safety of the ingredients is to look for signal words. If you see any of the following words, beware! The product is considered to be hazardous:

- caution
- warning
- danger
- poison
- flammable

- reactive
- corrosive
- toxic

If products with these labels are thrown in the trash, they can harm wildlife, soil, waterways, and humans alike. They must be disposed of at hazardous waste collection sites. If you or your center is in possession of these products, check with your waste hauler to learn how to properly dispose of them, and be sure to keep them locked up and well away from children.

Product labels include words that make promising claims too, but these do not always mean what you might think. Advertisers commonly use greenwashing, crafting labels that feature bucolic nature scenes or phrases such as "natural ingredients" or "green clean." But these images and phrases can be misleading. The following are a few examples.

Essential Oils

Essential oils are scented plant extracts often used in homemade cleaning products and sold for use in diffusers or air fresheners. Some make claims that essential oils can cure headaches and body aches, relieve fatigue, and improve mood. However, depending on their use and a person's sensitivity, some essential oils can irritate skin and mucous membranes. Approach essential oils with the same healthy skepticism that you would any product, and never apply them directly to a child's skin.

Nontoxic

Nontoxic implies that the ingredient is not poisonous or harmful. However, this term can be misleading because there is no standard definition for it as used to describe cleaning products. Even products labeled *nontoxic* can be harmful if used incorrectly, and they can cause harm to the environment even when used as directed. For these reasons, this term isn't helpful in making decisions about whether to use a particular product.

Organic

The word *organic* on a product label can mean just about anything, as there are no legal parameters that dictate when and how this word can be used. Products and foods in the United States are permitted to carry the US Department of Agriculture's "Certified Organic" label only if they are in compliance with the legal restraints established by the USDA, generally that there are no synthetic fertilizers or pesticides used in the growing or harvesting of plants.

Impact on the Planet

In addition to harming human health, many common chemicals are also problematic for the planet. Cleaning chemicals can harm wildlife and water quality, and some chemicals persist in drinking water supplies. High levels of certain chemicals originating in household cleaners have been detected in fish and animal tissues, drinking water, and ocean water (EPA 2009).

For these reasons, it's extremely important that you and your staff know how to dispose of products properly. Most cleaning products have disposal instructions written directly on the label. If the label doesn't have disposal instructions, look for the signal words from the preceding list, and contact your waste hauler for specific information.

Beyond chemicals in cleaning supplies, be aware that smog, vehicle emissions, and smoke can be harmful to infants and children, particularly those with asthma. This pollution can result in pneumonia, as well as other respiratory diseases and irritations (United Nations Environment Programme 2018). In 2017 the United Nations Children's Fund determined that some 17 million children around the world are breathing toxic air and that the vast majority of air pollution–related illness and death occurs in low- to middle-income areas, particularly urban areas (United Nations Children's Fund 2017). Air quality is definitely an issue of environmental justice and equity. Recalling the concept of critical windows of development, those harmful health effects can last into later life. If your program is located in an urban area that has a lot of traffic, you may want to keep the windows closed on smoggy days or purchase an air purifier to protect young lungs. Children's height makes them especially vulnerable to the effects of car emissions: their faces are close to tailpipe levels, which means that a walk along a line of idling cars at pickup time sends the toxins right into their bodies.

Takeaway Tip

Many of us grew comfortable with using masks to block particulates and aerosol particles during the COVID pandemic, and on days with poor air quality or smoke from wildfires, wearing a mask helps people breathe easier.

Health and Safety Considerations

- Always have the number of your local poison control office posted prominently. Of course, as you continue to replace more toxic products with those that are less toxic, the risk to children's health will diminish.

- Disinfecting wipes should never be used to clean children's hands or faces, even in a pinch. The chemicals used in commercial wipes are not meant for skin.

Practical Advice

- Know your local licensing requirements around cleaning and sanitizing.
- Spend some time reviewing your collection of cleaning products and researching alternatives.
- Use the websites provided here to seek out information on the products you can use to meet health and safety requirements while at the same time minimizing the health risks for the children in your care as well as the environment.

Involving Families

- Invite families from your community to visit your program and make Earth-friendly cleaners to take home.
- Create handouts or other informational materials to share with families so that they can understand hazardous waste and how to keep children safe from exposure.
- Reach out to your county environmental agency to find out if there are hazardous waste collection sites or special events near your program; then share that information with families so that they can dispose of hazardous waste properly.

More Information

For more information about the products and chemicals listed in this section, visit the US Environmental Protection Agency (www.epa.gov/children) and the Environmental Working Group (www.ewg.org). Both offer detailed, research-based, peer-reviewed information as well as additional research and resources specifically related to cleaning and personal care products. Additionally, your state or county extension office or environmental management office staff can explain where and how to dispose of hazardous waste and how to recycle containers that stored hazardous waste. Finally, check the EPA (www.epa.gov/greenerproducts) and Green Seal (www.greenseal.org) for specific product recommendations and additional details about so-called "green" cleaning products.

CHAPTER 3
REDUCING WASTE

Mr. Lopez knows that children learn the most by watching the important adults in their lives. For this reason, he makes sure his staff are constantly modeling pro-environment behavior such as recycling and reducing use of single-use plastics. He has developed an environmentally preferable purchasing plan for his business, and he ensures that all the paper products he buys are at least 50 percent recycled content. He and his staff reduce wasted plastic and other resources and are careful to teach children how to be thoughtful in their use of materials and supplies.

Mr. Lopez has the resources to frequently make positive environmental choices that might sometimes cost a little more, like buying recycled paper, but many of his actions cost no money at all, or in some cases, they may even save the program money. A program can take many steps to be more Earth friendly and reduce waste while at the same time modeling eco-friendly habits for young children.

Environmentally Preferable

What does it mean to be "environmentally preferable"? Simply put, if you have the choice between two products, the "environmentally preferable" choice has the least impact on human health and the environment. When thinking about this, you may take into account the signal words and ingredients described in the previous chapter, and you may also take into consideration the many factors that went into the creation of the product, for example, the materials, production, packaging, and even disposal of the product. Many program administrators and staff take the time to develop environmentally preferable purchasing plans, which serve as checklists of considerations for making purchasing decisions. See an example in appendix B.

Product Life Cycles

One of the luxuries of modern living in the Western world is easy access to products of all kinds. For those with means, most products—everything from clothing and household goods to art supplies, toys, puzzles, books, and more—are just one click away, delivered to your doorstep in a matter of days. But along with those goods comes a great deal of waste, especially packaging and shipping boxes. In fact, about 28 percent of all solid waste is made up of packaging alone! And many items wind up in the trash if they get broken or wear out. Only about 38 percent of non-food waste that is discarded in the United States is recycled (EPA, accessed 2021a). The rest winds up in landfills, where it literally piles up; it's sent to incinerators where it's burned, releasing potentially harmful chemicals into the air; or it's simply cast off into the environment, where we encounter it every day as litter.

It's easy to go about our days and not think much about how those goods flow in and out of our lives, much less the full life cycle of those products, from the raw materials used to create a product (inputs), to the ways in which a product is used, to the effects of creating, shipping, using, and disposing of those products (outputs).

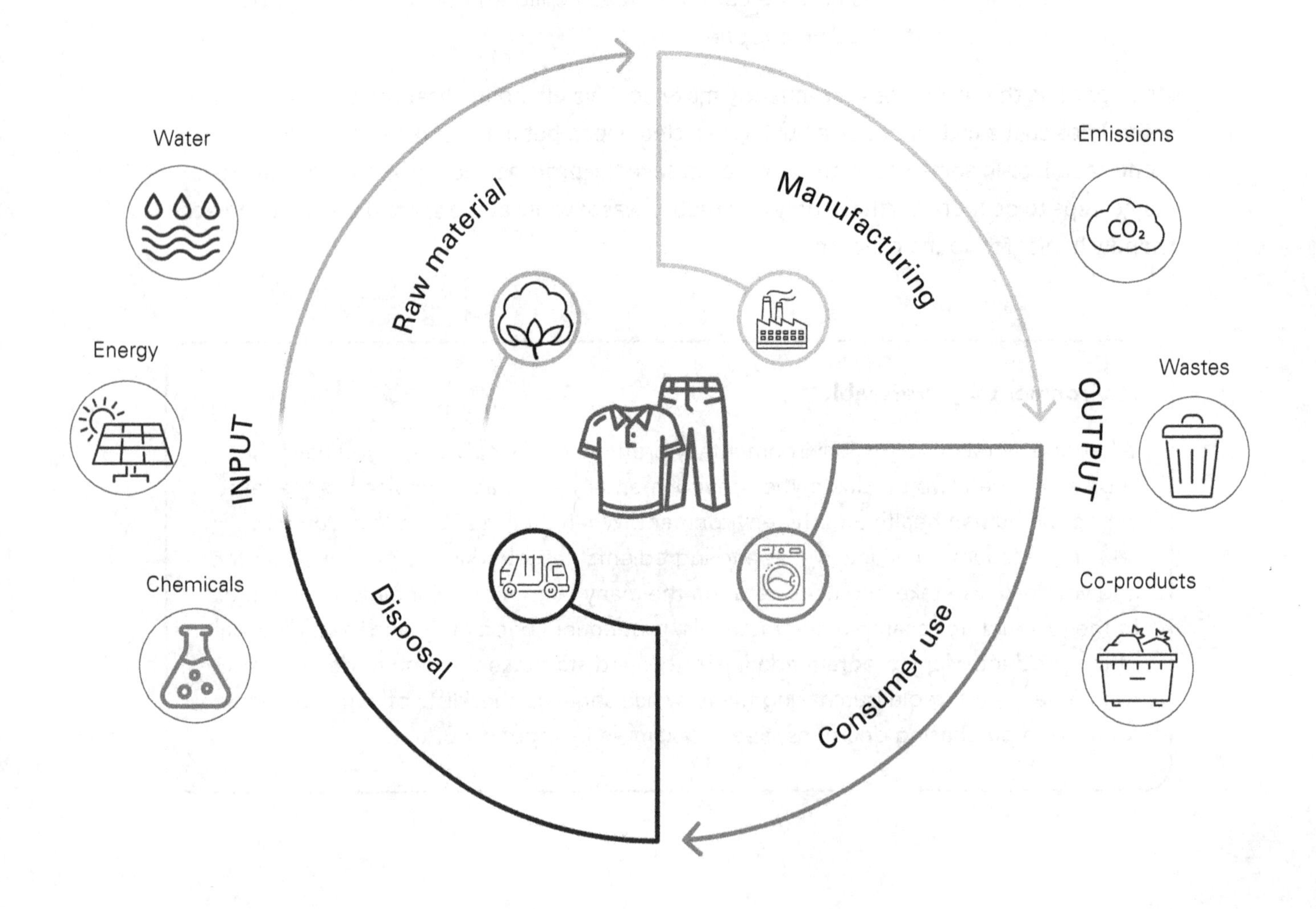

Looking at a product life cycle graphic, you can see the energy and raw materials used as well as the waste generated from everyday products we all rely on. This graphic doesn't even include the emissions involved in the shipping and delivery of goods. It also leaves out the intensive process of resource extraction to obtain materials: clear-cutting; use of oil, gas, and water; habitat destruction; and the human impact of the resource extraction process, as the workers involved in resource extraction often work for very low wages or have unsafe conditions. The US Environmental Protection Agency considers these additional factors when thinking about a product's life cycle: "ozone depletion, climate change, [ocean] acidification . . . smog formation, human health impacts, and ecotoxicity" (2021c).

Takeaway Tip

An important and impactful first step on the journey to a more Earth-friendly program is simply becoming more aware of the life cycles of the everyday products we consume. For example, it might take some effort to figure out specific information about how the brand of paper in your office printer was created and how it eventually wound up in your office, but a quick internet search can reveal some eye-opening general facts. Luckily, more and more vendors and suppliers are becoming more transparent about the sources, origins, and relative eco-friendliness of their products (just beware of greenwashing). More information about life-cycle analysis can be found at www.lifecycleinitiative.org or at www.footprintnetwork.org.

Reducing and Reusing in Your Program

The most important step anyone can take to be more Earth friendly is reducing consumption. This means buying fewer new products. When a purchase is necessary, it means avoiding nonrecyclable and single-use products when possible and instead selecting products that can be reused, recycled, or composted. In addition to generating less waste overall, reducing consumption of new products saves money. Depending on the choices you and your team make, you could save your program (and yourself!) lots of money.

While some licensing regulations do require programs to use disposable plates and single-serve containers, many programs choose this option simply for the sake of convenience. Check with your licensing representative to determine the requirements for your program. Then double-check. Sometimes licensors make mistakes or interpret the rules differently. If you are able to switch to reusable plates, cups, and utensils, you'll save money and resources.

What's more, the children in your program will learn to be gentle with dishes, and if your classroom arrangement allows it, they will enjoy helping clean up.

If you can't switch to reusable dishes, you can conserve at your site in other ways. If you work in the office, think before you print. Change your printer settings to print on both sides of the paper or consider sharing documents with your staff digitally instead of printing them. Email newsletters and menus to parents instead of printing paper copies. And there are many cases when it's perfectly appropriate to use a rag for cleaning (say, in staff offices or shared eating areas) instead of disposable wipes or paper towels. If you have the ability to launder items (or families willing to take on the job), cloth towels in the bathroom are much gentler on little hands than rough paper towels or noisy air dryers.

Reusing items also saves money and is more Earth friendly. Classroom materials and storage containers can often be used again and again, over time finding new purposes. Local thrift stores and reuse centers often have blocks, fabric samples, and other materials that make engaging loose parts for children's open-ended play. Families may be happy to donate materials that their own children have outgrown—don't be afraid to ask! Seeking donations of long-lived toys such as blocks, marbles, figurines, dolls, and other toys can save your program money. Asking families is also an excellent way to obtain outdoor gear such as coats and boots to allow the children in your program to safely and comfortably go outdoors to play. When you need to purchase an item, try secondhand stores before buying new. This keeps products out of landfills and conserves energy, and it's easier on your wallet.

Virtually every product we buy comes wrapped in plastic, cardboard, or other packaging material. Buying secondhand, buying supplies in bulk, and reducing your consumption all help reduce packaging and container waste. Packaging materials like bubble wrap offer many possibilities for the art area, and boxes are always a hit, especially in the dramatic play areas. Get into the habit of thinking about ways you can repurpose or reuse materials that you once would have discarded.

Takeaway Tip

While reducing consumption may not be as visible an action as setting up a fancy recycling center or hosting a schoolyard cleanup, it's one of the most important actions you can take toward becoming a more Earth-friendly program. The most sustainable change anyone can make is simply using less stuff overall, moving your program further from the extractive, consumptive product life-cycle process. What's more, you will reap the benefits too: research suggests that the personal impact of reduced consumption is more powerful than any other green action you might take (University of Arizona 2019). It offers a sense of well-being and security and increases your feeling that you have the power to create positive change.

Disadvantages of Recycling Plastic

Although efforts to recycle are usually beneficial, it's important to consider how you can reduce your use of everyday products—that way they don't have to be recycled *or* wind up as trash. Many practitioners can recall the early media hype around recycling: there were ads on city buses and television, and even the containers themselves touted messages encouraging folks to recycle them. Some cities and municipalities required that plastics, glass, and metals be washed and separated from one another. Other waste haulers allowed consumers to mix all the recyclable materials they generated. Consumers felt good, like they were doing something that had a real impact on the environment. The trouble is, more often than not, materials we thought were headed for a new life as a recycled product simply wound up in landfills. In order for there to be a demand for anything, there must be a financial market for the goods, and this applies to recyclable materials just as it does to other materials. In some parts of the country there is a demand for certain types of plastics but not others, while other parts of the country recycle only one kind. These demands fluctuate and vary widely from one locale to another, causing confusion about what can and cannot be recycled. In addition, consumers have been misled since the 1970s about the reuse potential of plastics, but plastic degrades every time it's reused for any reason. Recent news reports have revealed that many of the messages earlier generations received about recycling were in fact promoted by plastic manufacturers to create a consumer demand for plastic (Sullivan 2020). These companies knew that there would be no viable, sustainable strategy for the recycling and reuse of plastics, but they put profits ahead of ethics. The result? We are now in a situation where there is so much wasted plastic that it is clogging our waterways and oceans. Although there are still places where plastic is recycled effectively and efficiently, the far better choice is to avoid plastics.

Health and Safety Considerations

- Children with food allergies can have reactions to residues left behind on food containers that aren't properly cleaned or sterilized. For this reason, it's imperative that you wash and sterilize all food containers before reusing them.

- Different communities have varying rules about what can be recycled. Find out what rules apply where you are located by contacting your local government's website, your county extension office, or your waste management company.

- Know the rules and regulations that apply to your program surrounding food and packaging. Can you buy food in bulk, or does food need to be prepackaged and individually wrapped?

Practical Advice

- Most early childhood educators are already experts in creative reuse. Look around your own classroom and chances are you'll see lots of repurposed materials, from plastic water bottles to paper towel tubes, fabric scraps, and more. You are probably already reusing and repurposing a great many items. Way to go! You'll still find some new ideas throughout the pages of this book.

- Children are easy targets for advertisers. When you hear young children talking about the latest and greatest toys, elicit their opinions about why the toys are so popular. Facilitate a discussion about needs versus wants.

- Research the cost-to-benefit ratio of buying in bulk. You will be surprised at how much your program will save over buying smaller (or single-size) portions of food.

- If your program is required to use single-serve, individually wrapped foods at snack and mealtimes, seek products that come in containers you can reuse or recycle.

- Model behaviors of reducing consumption. Can you think of a time when you wanted something but decided not to buy it? Share stories with children and their families of your own journey toward becoming more Earth friendly.

Involve Families

- Have a clothing drive at your school to promote reuse. Invite families to bring clothing their children have outgrown to pass down to smaller children. This builds community in your program while helping others.

- Toy swaps have similar benefits to clothing drives. Families bring in toys that are in working condition but their children have outgrown or lost interest in. Then they pick new toys from the donated ones. Keep toys in circulation for a long time by creating a "toy library" that families can visit when it's time to refresh their children's play options.

- Offer a parent education evening on the benefits of reducing consumption. Many families are looking for ways to save money and will appreciate tips on how to reduce purchases and repurpose items.

- If your program allows, use cloth towels and napkins instead of paper. If you don't have a washing machine available, recruit families who are willing to launder and return items. Provide them with Earth-friendly detergent if possible.

- Before you buy new, reach out to your community. Ask parents to donate items you need that are in good condition.

More Information

Learn more about product life cycles and reducing consumption via the Story of Stuff project, www.storyofstuff.org. This website offers resources for learning where products come from and how we can reduce our use of many things. It also has short, informative, family-friendly videos that explain many of the concepts discussed in this chapter.

Visit each of the websites listed here to access a bevy of resources, information, and child-friendly activities related to creatively and carefully disposing of items, reducing consumption, and repurposing things instead of throwing them away. Find information about what and how to recycle common products like juice boxes, food wrapping, and much more at http://earth911.com. The site also has craft activities and fun facts to explore. The Center for a New American Dream website (http://newdream.org) helps redefine our consumer culture. It publishes news related to consumer issues and resources for those looking to reduce their carbon footprint.

CHAPTER 4
EARTH-FRIENDLY FOOD AND GARDENING

Mr. Lopez has invested time and money into creating a community garden on-site. In the region where his center is located, the growing season is long, allowing ample time for vegetables to grow. The teachers on staff have all taken an interest in using the garden as their classroom and practice organic gardening, avoiding the use of pesticides or chemical fertilizers, harvesting and storing rainwater to keep the gardens from getting too thirsty, and involving the children in caring for the garden. The children are eager to get their hands dirty, and they enjoy the tasty, fresh vegetables at snacktime. They are appreciative of the vegetables they grow themselves and because of this are careful not to waste.

Not everyone can start a community garden, but you can evaluate your program's planning, buying, cooking, and waste practices and make the changes that feel manageable for you and your community of families.

Sustainable Food Choices

Mealtimes and snacktimes are highlights of many days in the early childhood classroom. They are social activities that build community and help children learn life skills, practice language, math, and literacy, and of course, make new friends. One of the most Earth-friendly things a teacher can do is to examine their program's food practices and make shifts toward sustainability, as this approach contributes to the health of children *and* the Earth. Foods that are minimally processed are best because they are free of artificial preservatives, flavorings, and colorings. You can keep this helpful guideline in mind when choosing and serving high-quality and sustainably produced food.

In today's busy world, families often scramble to make the time for a meal at the end of each day. Have you ever had a child in your program who was a "picky eater"? Or one whose caregivers assured you they would eat only certain foods? Barriers like these lead many caregivers to reach for processed foods: quick, packaged meals like macaroni and cheese, pizza, or canned spaghetti—things they know children will eat. Snack foods like french fries and chips are quick and can be consumed on the go. Common in many early childhood settings are processed snack foods like cookies, crackers, breads, and cereal bars. Part of the reason why these foods are so pleasing to the palate is because they contain high levels of sugars, sodium, and fats—all flavors and qualities that are appealing to a wide range of eaters. Many contain preservatives and other chemical ingredients designed to lengthen the product's shelf life, retard mold growth, inhibit bacteria, and add extra sweetness or color. They are often high in calories but low in nutrients. Processed foods can make a person feel full because they're absorbed quickly into the bloodstream. Unfortunately, that feeling quickly fades away once those calories are burned, leaving the person feeling hungry, jittery, or cranky. Moreover, diets rich in processed food tend to be associated with poor health outcomes, such as obesity, high blood sugar, diabetes, and digestive disorders. Children who consume lots of processed foods are more prone to these poor health outcomes as they age, meaning that a nutritionally deficient diet in childhood can have long-term effects.

More nutritious options, like whole grains, fresh vegetables, and legumes are much better for children's growing bodies because they are high in vitamins and minerals. They also take longer to digest, meaning that children feel full longer and are less prone to the mood swings and cranky behavior often associated with the blood sugar spike that accompanies processed foods (Fuhrman 2018).

Clean Plates and Food Waste

Are you a member of the "clean plate club"? Many people grew up in households where eating everything on one's plate was the norm, as a way of reducing food waste. While reducing food waste is a laudable goal, forcing children to clean their plates can lead to conflict and unhealthy relationships with food. Allow children to stop eating when they feel full. If there is extra food on their plates, add it to the classroom compost bin for the garden.

As food is being served, ask children to "check in with your body" or "ask your tummy how much food to take" so they can start to develop a sense for what hunger and fullness feel like to them. Never shame a child for taking too much. Some children tend to hoard food or take large amounts of food in response to traumatic events or situations at home, such as food insecurity.

The miles your food travels from farm to fork are a large factor in determining its sustainability. Locally grown foods require far fewer resources in transit to your plate than foods grown across the country or across an ocean, considering that transportation requires gas, oil, and water, as well as carrying a human cost. This makes local food a sustainable choice, significantly reducing negative environmental outputs. What's more, in many cases, locally grown foods will use less water, since smaller farmers (who in many regions happen to be local growers) often choose crops that are better suited to grow in their particular climate. Note that in some locales, orchards, croplands, and farms stretch as far as the eye can see and are often very much *not* sustainable—this is one more reason to know where your food comes from. For example, if you have the option of buying raspberries from a small local farm or from a large-scale agricultural empire an equal distance away, consider which entity has more resource-extractive practices. Which farm uses more potentially harmful chemicals? Thinking from a justice perspective, which farm treats its workers more fairly?

Consider factors such as whether food is organic, which means that in its growth and production fewer (if any) chemicals are used. Often organic food is more costly than conventional, and some state-funded nutrition programs don't support purchasing organic. These factors can make providing organic lunches or snacks cost prohibitive. But check the rules in your area. The USDA label Certified Organic can be considered the gold standard, as certification carries with it adherence to very specific rules and regulations and requires that farms have been free of pesticides for a certain number of years. This helps to explain the high price tag of organic foods. Know there are many local farmers who avoid chemical use and may be working toward organic certification but aren't there yet. You can often find local growers at farmers markets who practice Earth-friendly farming and growing techniques, even if they are not certified organic, and because you meet the farmer at the market you can personally ask about their farming practices. Another strategy that might make sense for your program is joining the community-supported agriculture (CSA) movement. Many small farms throughout the United States offer memberships similar to subscriptions, with customers receiving a box of fresh produce from the farm on a weekly, biweekly, or monthly basis. This sort of arrangement allows you to support small farmers, buy locally, expose children to a variety of healthy fruits and vegetables, and receive abundant (and in many cases organic) foods to enjoy together. To find out more about how CSA programs work in your area, visit local food co-ops where these small farms often advertise. You can also check with your local county extension office, do an internet search for "community-supported agriculture near me," or check out the US Department of Agriculture's page about CSA programs (www.nal.usda.gov/legacy/afsic/community-supported-agriculture), which includes a searchable database.

Livestock production contributes over 14 percent of the total worldwide carbon emissions (FAO 2013). The industry relies heavily on fossil fuels and water use, and meat tends to be more expensive than other types of food. For this reason, giving up meat has been identified as one of the most impactful things a person can do on behalf of the environment (Panoff 2021). The vast majority of Earth's population relies on whole grains and legumes for protein—it's only the wealthier, more polluting nations that rely on meat for so much of our food consumption. This is an issue of environmental justice: Western countries, which have normalized meat at practically every meal, drive the meat and factory farming industry, but the environmental effects of this practice are felt around the world.

Plant-based diets can offer long-term health benefits, such as healthier hearts, better circulation, appropriate weight, lowered risk of disease, and more (Hever 2016). Many healthy whole grains and legumes are good sources of protein. Readily available, inexpensive, and adaptable, rice and beans are a staple in many countries around the world, and by increasing your program's consumption of rice and beans over meat, you not only save money but also increase your and your students' cultural sensitivity. Check out the sites in the More Information section to access recipes. You will also find tips for education and research so you can involve families in this positive change.

Practitioners who aren't comfortable with the idea of giving up meat in their programs altogether may be more willing to try "Meatless Mondays." Started by the Johns Hopkins School of Medicine, the practice of Meatless Mondays was established to encourage organizations to avoid meat. By giving up meat just one day per week, people can reduce their environmental footprint, improve health outcomes, and even save money.

Takeaway Tip

Recently, companies have begun labeling certain foods as "plant-based," but this is a form of greenwashing. Ultimately, all food is plant-based—even meat, since farm animals eat plants before they are processed and turned into meat. And just because something is "plant-based" doesn't mean it is free of chemicals or artificial additives. This is another good reason to read labels carefully!

Gardens and Compost

Creating a school garden encourages children to eat healthier, take care of the Earth, and have fun outdoors. Even if you don't have a lot of space to work with, you might be able to find room for a container or two, such as a bucket or window box, where you can plant herbs or other easy-care vegetables, many of which require little more than soil, sunshine, and water. For beginning gardeners, I recommend purchasing seedlings or starter plants to plant outdoors once the weather is warm enough rather than growing plants from seeds. While planting seeds can be an excellent learning opportunity under the right conditions, in many parts of the country the growing season is too short for many plants to reach maturity if they are directly seeded into the ground. If you're so inclined, starting seeds indoors can be a powerful learning experience for children and teachers alike. Factor in considerations like types of vegetables, growing zone, and amount of sunlight, and once you've done a little bit of research, you can get started on what will surely be an ongoing source of excitement in your classroom. An avid gardener from your family community would be a great resource in selecting plants that will thrive in your area, or check with your county extension program to find master gardeners who can help you.

Many herbs and vegetables are quite happy growing in containers. I recommend basil, dill, cilantro, chives, and mint. Also easy to grow in containers are bell peppers, black beans, zucchini, green beans, tomatoes, and peas. Depending on your site, you may be able to incorporate pumpkins or squash—they grow readily in most sunny conditions, and the flavors are generally pretty agreeable to children. Of course, there are many factors that matter: your site, access to water, the environmental conditions (Sunny or shady? Length of growing season?), how much your colleagues will commit to helping you, and your own know-how and willingness to go outdoors to tend a garden, even in uncomfortable weather and on weekends.

Whatever you can do—do it! Children love growing their own food, and it can help create healthy food habits. In addition, gardening is interdisciplinary and fosters lots of academic connections, and it encourages you and your students to be outdoors. Eating food you've grown together can build community, expose children to different cultural practices, and create opportunities for connection and social-emotional development. Taking care of plants and ensuring they have what they need—sunshine, soil, and water—are steps toward taking care of the planet. When children have the opportunity to take care of their immediate environment, including the material objects as well as living creatures such as animals and plants, they are more inclined to care for the larger environment overall. Most of all, having a garden permits you and the children to enjoy the fruits (and vegetables!) of your labor: pass around tomato slices and fresh chopped herbs at snacktime, pick peas off the vine and enjoy them in the spring sunshine, or sauté bell peppers to top your beans and rice at lunchtime!

Takeaway Tip

Reduce food waste by incorporating a composting program in your classroom. If you have outdoor space or a garden, the compost is a useful addition and will help your plants grow and be healthy. For beginner composting tips, see www.epa.gov/recycle/composting-home. A barrel or bin won't take up much room at all, and it won't get full that quickly if you keep tumbling it, as the material keeps breaking down into smaller pieces.

Health and Safety Considerations

- Be aware of the allergies of the children in the program before you do any planting. Many children who have peanut allergies react to peanut plants and may also react to pea plants, whether through direct contact or ingestion. Be sure you are aware of the severity of any child's allergy.
- If you can't transition to serving all or many organic foods, become familiar with the Environmental Working Group's "Dirty Dozen" list of the top twelve fruits and vegetables that when grown conventionally tend to have the highest pesticide loads. Starting with the dirtiest, they are strawberries, spinach, kale/collard and mustard greens, nectarines, apples, grapes, cherries, peaches, pears, bell and hot peppers, celery, and tomatoes (EWG 2021).

Practical Applications

- Encourage children to talk about their food during snack or mealtimes. Help them learn about ingredients and where food comes from. Share stories about favorite foods and special meals.
- If you purchase food for the classroom, research the sources of the food you buy. Is it locally grown? How far did it travel to get to your classroom? Is there a local option that might save you money?
- Consider keeping a worm bin in the classroom. Classroom composting systems are easy to keep indoors and offer year-round opportunities to care for "pet worms" who convert leftover food scraps into compost you can use in your classroom plants or outdoor garden. This is a fun and easy way to reduce leftover food waste. This approach has many educational benefits as well: from learning about life cycles of worms to the process of decomposition, classroom composting can inspire lots of projects and investigations that your students will enjoy.

Involving Families

- If your program space allows for an outdoor garden bed, invite families to volunteer to take care of the garden on a regular schedule. This can build community.
- Invite families to share favorite recipes at a recipe swap, or consider hosting a potluck and invite families to bring some of their favorite dishes to share.
- If you make specific recipes in the classroom, send home pictures of children enjoying their creations along with the recipe so they can "teach" their families how to make their favorites from school.

More Information

To find garden activities and garden-based lesson plans for kids, as well as information about grants for early care and education programs seeking to start or expand a garden, go to https://kidsgardening.org. Check out the pages specifically about seed starting indoors: https://kidsgardening.org/gardening-basics-indoor-seed-starting-qa. At www.nrdc.org/stories/composting-101 find a composting guide for the home gardener explaining several composting processes.

The US Department of Agriculture has a website (www.fns.usda.gov/tn/standardized-recipes-cacfp) with recipes and menu planners specifically meant for early care and education programs. Lots of creative, healthy, and culturally responsive recipes can be found here, many of which are meatless.

CHAPTER 5
EARTH-FRIENDLY PLAY

Mr. Lopez has thoughtfully considered play materials for his classrooms to ensure they reflect an Earth-friendly mindset. He buys toys and materials made from recycled materials whenever possible and has thought about how to dispose of items when they become broken or worn out. He supplies his classrooms with nontoxic paint and glues, and rather than stocking the dramatic play area with plastic props, he has furnished it with pans, trays, and utensils purchased from secondhand stores and yard/garage sales that he sanitized carefully. He also provides toys and books that help children directly connect with the natural world, featuring animals, plants, and settings the children will recognize.

By reflecting on the origins of the toys and materials in his classroom, Mr. Lopez is not only saving money but also modeling for the children and families how easy it can be to use secondhand and recycled toys and art materials every day. He is also demonstrating that these items still hold a lot of value.

Sustainable Play

The heart of any early childhood classroom is play. As an educator committed to children's growth, development, and learning, you already know the many benefits of play in the early childhood years. Taking a few steps now can ensure that your program's play materials are sustainable, safe, and ready for lots of Earth-friendly play for years to come.

Is it strange to think of a simple toy as potentially compromising the ability of future generations to meet their own needs? While a plastic car may seem pretty harmless, consider its life cycle (as illustrated in chapter 3) and all the resources that went into the production, manufacturing, shipping, and distribution of that toy, not to mention what happens to it at the end of its useful life. When we think about sustainable products, we consider all of those steps in the process and consider the overall impact of the product. At first, thinking about whether toys and play materials are sustainable might feel overwhelming. It's understandable since you probably have lots of different kinds of toys and lots of different places to play. This chapter identifies some of the more commonly used playthings in a classroom setting—sand and water, art supplies, dramatic play props, musical instruments, and building materials—and gives you tips for finding Earth-friendly, sustainable options or substitutions.

Sand and Water Play

Young children love sand. It's soothing to the touch, quiet, and endlessly fun for burying things as well as pouring, dumping, and moving from container to container. Many programs choose not to have sand due to the slipping hazard, while others have it outdoors only, but sand makes an excellent loose part in nature play or at a sensory table. Sand is a better choice for a sensory table than food such as rice, corn, or beans. In addition to being wasteful, using food in sensory tables can be triggering for children with food insecurity, and it can also present a risk of allergic reactions. Cultural sensitivity demands that practitioners avoid using food in sensory tables.

Takeaway Tip

Sand isn't a renewable resource, but it can be reused once your class is through with the sand table play. Add sand into the garden bed, sprinkle it on icy sidewalks, fill socks for beanbag toys, mix it with paint or glue to make an art medium, fill terrariums, or top off plant pots.

Water play is another childhood favorite. We are fortunate in much of the United States to have an abundance of water, and early childhood educators can attest to the endless play possibilities of water. But some areas are drought prone, and clean water truly is a luxury. Water that is not toxic to humans but not potable (consumable) is called gray water. Water used for rinsing toys or art supplies, rainwater collected in buckets, or water leftover from art projects such as watercolor painting—all of this is considered gray water. Gray water can be used for watering plants, washing outdoor toys and equipment, and making sand and mud for outdoor play and outdoor art projects. Collect watercolor painting water to water plants or save it for "painting" on sidewalks outdoors. This sort of use is much less wasteful than simply letting that water go down the drain, and it models treating water like the precious resource it truly is. For safety and to avoid spreading germs, don't reuse water from washing mealtime dishes or water that is leftover in water bottles.

Rainy days are perfect for going outdoors and playing in the water as well as jumping in puddles. Take children to explore the environment surrounding the program to discover where water goes during a storm. Where do the puddles form? Where does the water soak in? Is there lots of green space (such as gardens or other places with plants) where the water will be absorbed, or will it run off, finding its way to the nearest stream? How can you find out? Can you collect water during a storm? If so, what can you do with it later? How can you collect it? What spots catch the most rain (likely the open areas), and which ones catch the least? (under trees or an awning). These explorations can be just as exciting as the classic "sink or float" and pouring games so popular at classroom water tables, and they make use of water in a way that is more Earth friendly and less wasteful. Beyond that, and perhaps more important, they encourage children to get outdoors and play regardless of weather conditions. This gives them the opportunity to explore the environment and learn more about where they live.

Art Supplies

Art projects present an opportunity for reusing materials rather than tossing them into the recycle or trash bin. You probably have lots of materials for collages, containers for decorating, cardboard and paper for cutting, fabric scraps for sewing, and so on.

In addition, consider the ingredients and production of the art supplies you purchase. For example, did you know that commonly used art supplies can contain lead? Remembering the signal words from pages 14–15, review the ingredients in your paints, markers, chalks, and glues to determine how safe the products are. Companies aren't required to list all the ingredients in these products, so you may have to do a bit of internet sleuthing to know whether your products have toxic ingredients or nonrenewable resources.

In response to the growing interest in sustainability, many companies have put a new spin on their products to make them appear healthier or less wasteful, described by the term *greenwashing*. For example, a product label might state that the product is "made with recycled materials" when just a small percentage of the raw materials are recycled. Also, many products, such as plastic bags, are labeled "recyclable" even though they are made from a type of plastic that is unacceptable in most recycling programs.

Many small craft supplies like googly eyes, pom-poms, and plastic buttons are not recyclable, so use them in limited quantities and reuse them whenever possible. Glitter, while sparkly and fun, is impossible to reuse, and the tiny pieces easily get into the environment, washing down the drain and entering waterways. Most glitter is actually a combination of plastic and aluminum, meaning that it takes a very long time to biodegrade. Once it enters rivers, lakes, and the oceans, it may be consumed by fish or other wildlife, and it can accumulate as microplastics in the oceans.

Homemade Paint

The internet is full of recipes for homemade paint, but many of them contain harsh detergents, creams, or other personal care products that are hazardous for young children. Here are three easy and safe recipes:

1. Combine 4 tablespoons of baking soda, 2 teaspoons of white vinegar, ½ teaspoon of light corn syrup, 2 tablespoons of cornstarch, and food coloring. You can add more or less food coloring as desired. This makes for a fun color-mixing activity, but be aware that food coloring will stain hands and clothing.

2. Gather up dried-out markers and soak them (by color) in tubs of warm water. The water takes on a light hue and makes a fine watercolor paint.

3. Mix equal parts flour, salt, and water for a thin paint you can pour into squirt bottles. Add food coloring, and then shake the ingredients together.

Don't overlook the potential of natural materials for creating pigments and textures. Mud comes in a wide range of colors depending on the weather conditions, and soils come in an array of earthy hues. Grasses and leaves, when crushed, also leave color behind. Go on a hike and explore all the art materials and colors to be found in nature.

Dramatic Play Props and Musical Instruments

Dramatic play allows young children to practice behaviors and mimic adults, and stocking this area with thoughtfully selected Earth-friendly materials helps instill an environmental mindset in children. There is no shortage of materials that can be obtained secondhand and added to the dress-up and dramatic play areas. Your dramatic play area provides an opportunity to remove single-use plastics: as plastic toy cooking equipment breaks or wears out, can you replace it with secondhand pots and pans from the thrift store? Can you find jewelry, clothing items, photographs, or prints to liven up the space?

Plastic musical instruments are often poor quality and break easily. The plastic they are made from is rarely recyclable, so choose wooden or other natural materials such as coconuts, gourds, and seedpods as you fill out your instrument collection. Making homemade instruments reuses containers, paper tubes, and other craft materials.

Building Materials

Blocks and other construction materials are always popular in the early childhood classroom. At the time of this writing, most plastic building materials are nonrecyclable, which means that despite their popularity, they aren't the most Earth-friendly item. They can be sanitized and used for years, which makes them a reasonably good choice in terms of durability and longevity: you certainly won't have to replace them year after year. On the other hand, blocks made of wood or from recycled materials can also be a good choice. Many companies, responding to consumer demand, are taking steps to be more sustainable in the sourcing, production, and packaging of their products. But you don't need to purchase brand-new, often expensive toys. Consider visiting thrift stores and using social media to find low-cost or free materials.

Health and Safety Considerations

- Do your research to avoid getting hooked by unsubstantiated claims. Before investing in products for your classroom, do a little digging online. You can often learn more about products from the website than you can from a label or a brief description on the packaging itself.
- Seek out product reviews from consumer protection websites, such as the Consumer Product Safety Commission (www.cpsc.gov), whose function is to inform consumers about product recalls and other health and safety issues that emerge.

Practical Applications

- If your site is near a natural area, go on a treasure hunt with children. Venture out and collect acorns, stones, sticks, seedpods, or other natural materials that can be used as loose parts that can take on any role in children's play. Often, loose parts collected from natural areas are appealing art materials. Have you ever painted with pine boughs, used mud as a medium, or stamped with paint using stones?
- You may already use homemade modeling clay or dough. The internet has many recipes for scented or colored modeling clay, so consider branching out from your standard recipe. Be aware that most recipes contain wheat flour, so if you have children with gluten intolerances in your program, it's best to avoid this project.
- After thoroughly cleaning and sanitizing them, cut up old sponges into fun shapes to add to your art materials bin. Children will love painting with them, using them as stamps to create patterns, tracing around the shapes, and stacking them up.

- Rather than throwing away nonrecyclable plastic such as empty yogurt or applesauce containers, fill them with small stones, sand, bells, or beads. Cover with a fabric scrap and secure with ribbon or a rubber band to make instruments or sensory play toys.

Involving Families

- In need of more toys or other classroom materials but don't have the time to go thrifting? Consider building a "wish list" with your specifications, and then ask the families in your program to donate items they no longer use. This reduces consumption, reuses materials, and involves families in your efforts.
- Invite families to contribute to the classroom library by sharing favorite storybooks. This can help build a classroom library with a diversity of characters, places, and cultures portrayed. Representation matters to all students, and it's good for all children to see children of all races and backgrounds, family structures, and identities.
- Families are often eager to share culturally specific toys, musical instruments, or even clothing with their child's classroom. If this is the case, extend an optional invitation to the child or their family to come for a visit and teach the class about their items. But be mindful to avoid cultural appropriation, surface-level education, or "cultural tourism." And never press a child or family to share about their culture if they haven't expressed interest in doing so.

More Information

While you can research and learn more about most products online, many manufacturers use greenwashing to convince consumers their products are safe, nontoxic, and Earth friendly. The best place to go for art supplies information is the Art and Creative Materials Institute (https://acmiart.org/). This organization regularly reviews commonly used art supplies for safety. You can look up products by name using its certified products list and find resources related to toxicity and safety.

There are places to recycle some toys. For example, some toy manufacturers accept toys to be recycled. In some cases, the toys are cleaned or refurbished and sent to schools around the country. Some companies will take their products back just for the cost of shipping them. At the time of this writing, Crayola takes dried out markers from any manufacturer (though the program was paused during the pandemic), crayons can be returned to www.crazycrayons.com/recycle-program, and Lego takes old Lego bricks (www.lego.com/en-us/sustainability/environment/replay). Toys that are still in working condition can be donated to domestic violence shelters or other community organizations.

The following websites all offer curriculum, activities, and background information on water, water quality, and water-related issues for teachers. Start here as you think about how to make shifts toward being more water aware. The Water Sense website from the Environmental Protection Agency (www.epa.gov/WaterSense) offers water-related information, activities, and projects for educators. Project Wet (www.projectwet.org) is a national environmental education curriculum, and the website contains resources focused on water education, including resources specifically for early childhood. Water Footprint (www.waterfootprint.org) helps you calculate the water cost of your consumer habits. It includes materials in English and Spanish.

CHAPTER 6
INCORPORATING NATURE INTO THE LEARNING ENVIRONMENT

Mr. Lopez values the outdoors and wants the children in his program to feel the same way. He is lucky to have many windows in every classroom, and he has made sure that despite his school's urban setting, outside each window is a scene with green plants, flowers, and bird feeders. He knows that a view of the outdoors is beneficial for children and adults. He has opted to use natural, soothing materials throughout the classroom, like unpainted wood, natural fabrics, and loose parts from nature such as pine cones and shells for play. The teachers and children make frequent visits to the school garden, where they stop to check on the vegetables they have planted and where they watch butterflies and ladybugs make their rounds. Mr. Lopez supports his staff in taking children outdoors for play throughout the day so that children can connect with nature and be inspired to care for the Earth.

Mr. Lopez's setting may sound like something out of a fairy tale, but incorporating nature into learning spaces isn't out of reach for any teacher. This chapter addresses some reasons that teachers choose to incorporate nature, including the potential benefits to children's development and learning.

Learning in and from Nature

The term *nature* means different things to different people. For some it means wilderness areas far from the influence of humans. For others it might be a dandelion growing through a crack in the sidewalk. Nature is everywhere, and there are plenty of ways and places to enjoy it, learn about it, and help children be inspired to care for it. Many educators who wish to be more eco-friendly see a clear benefit to using the natural environment as a place for learning. Indeed, research shows that children who spend time outdoors with a caring adult grow up to be people who care for the environment. Simply knowing about or experiencing nature is just one piece of the puzzle—children need to have strong positive feelings about the Earth too (Kuo, Barnes, and Jordan 2019). What that means in the early childhood years is that in addition to learning about the Earth, children need to spend lots of time exploring and feeling safe and joyful in natural settings and with natural materials.

Time in nature has been shown to decrease stress and anxiety in children and adults, and who doesn't need that? When they have time to engage in free play in nature, young children have shown increases in creativity and problem-solving skills (Moore 2014). Playing in nature offers many opportunities for working together, which is good for children and their developing social skills.

Along with plenty of outdoor play, teachers can help young children engage with the Earth and animals by filling bird feeders, planting flowers or vegetables, watering plants, and doing crafts outdoors. Gardening allows children to connect with nature, learn about plants and animals, and enjoy learning about healthy foods, to boot. The natural world is rich with learning opportunities, particularly those associated with science, technology, engineering, and math. For example, children engage in science through firsthand exposure to life cycles in nature. They use technology when they select tools for digging, carrying, dumping, and moving materials outdoors. Building structures out of sand, snow, or sticks engages them in engineering. And math is happening everywhere as children make patterns, measure and stack materials, share and combine their play objects, and move their bodies over, under, around, and through natural spaces. Plenty of research supports teaching outdoors for its academic benefits, and being outdoors also boosts cooperation, inter- and intrapersonal skills, and social-emotional learning. Finally, in response to the COVID-19 pandemic, many teachers began spending more time outdoors to reduce virus transmission risks through aerosolized particles and to allow for more physical distance.

Here is a list of tips to help build going outdoors into your daily routine:

- Make it clear in your caregiver handbook that your class will be spending time outdoors every day. Include a contract or permission slip that caregivers sign granting permission to do so.
- Ensure you have spare outdoor gear on hand so that all children can comfortably and safely participate in outdoor activities. Depending on where you are located and the season, this may include warm layers, rain gear, sun hats or warm winter hats, mittens, boots, bug repellant, and sunscreen.
- Talk with your colleagues and administrative team to develop an emergency plan. This should include carrying two-way radios and first aid kits and planning where to meet should the groups become separated.
- Natural Start offers a free Professional Practices Guidebook (https://naturalstart.org/nature-based-preschool-professional-practice-guidebook) filled with safety tips, clothing

suggestions, planning recommendations, and curriculum ideas for teachers who want to take their classes outdoors. It is appropriate for seasoned outdoor teachers as well as those just getting started.

Admittedly, going outdoors every day is just not practical in some settings. The weather isn't always conducive to outdoor adventuring, there may be liability concerns, children may not have the right gear to be comfortable and safe, and there will be days when you just aren't up for it. In many cases these issues can be reduced through engaging with and educating families and your administration.

You can still help foster children's love for nature by bringing the outdoors in, for example, offering sticks, pine cones, acorns, small stones, or shells—loose parts found in nature that can be used for art materials, math manipulatives, alternatives to plastic figures, and props in many dramatic play adventures. They work well for learning and practicing important math skills like patterning, counting, ordering, and classifying. Another benefit to using nature-based materials in the classroom is that they are always biodegradable, they can be reused again and again, and the life cycle of natural materials is a science lesson unto itself.

Takeaway Tip

If you're short on green space, consider adding some lush houseplants to the classroom space for a more natural feeling. Place bird feeders, pollinator-friendly plants, wind chimes, suncatchers, and mobiles outside classroom windows so children can look out and enjoy nature anytime. Check the American Society for the Prevention of Cruelty to Animals website to learn which plants are safe to have in classrooms where animals may be present: www.aspca.org/pet-care/animal-poison-control/toxic-and-non-toxic-plants. The website for the US Department of Health and Human Services Office of Head Start has a list of plants that are poisonous and best avoided in early childhood settings: https://eclkc.ohs.acf.hhs.gov/safety-practices/article/even-plants-can-be-poisonous.

Remember, the most Earth-friendly approach is to reduce consumption. With the rise in popularity of eco-friendly settings and nature-based learning, there has been tremendous growth in the supply of green products designed to make your setting look and feel more Earth-friendly. Getting rid of unsightly, bright-colored plastic in favor of a more natural feel can be tempting. But if you have plastic bins doing a fine (but maybe not aesthetically pleasing) job of storing your classroom toys, resist the urge to get rid of them only to buy a set of new eco-friendly woven baskets.

Finally, even if you don't have a lot of green space outside the classroom, you can incorporate an Earth-friendly mindset by taking your previously indoor activities outside. Is there a shady spot where you could have a picnic at snacktime? Could you take a walk around the neighborhood to look for birds? A demonstrated sense of curiosity and wonder toward nature will have an enormous impact on the children in your care.

Health and Safety Considerations

- Before collecting loose parts outdoors for your classroom, be sure you are aware of any known allergies or sensitivities. For example, children who are allergic to tree nuts could react to handling walnuts or acorns, so it's best to avoid bringing them to the classroom.
- Many natural materials can be collected from parks, beaches, and other public spaces for use in your classroom. Consult with the appropriate authorities to be sure it is legal to take items away.
- Be sure that all children in your program have proper gear for going outdoors. A "gear swap" with families and community members can help make sure that everyone has a rain jacket, boots, and other weather-related gear so they can stay safe and comfortable outdoors.

Practical Applications

- Whenever possible, try to spend at least *some* of your outdoor time away from play structures. While those colorful structures are enticing, children love to play and explore in other areas as well. Don't have another contained space? How about a walk around the neighborhood to look for interesting trees or birds?
- When the time comes to update your play space, consider shifting to more of a nature-based setting, complete with small hills to play on, trees and other sensory plants to enjoy, natural fencing, and open areas to run and play. Many resources on creating nature-based play areas for early childhood can be found at https://naturalstart.org/professional-development.
- Many routine classroom practices can be taken outdoors. To get started, ease into things: once a week, have snack or story time outside on a blanket.
- Invite every child to bring in an object from nature to share with the class to build your library of nature-based materials.
- Many teachers report that starting their day with outside time can increase concentration and cooperation indoors. This might be due to the amount of social activity and practice at self-regulation that children have when they are given the chance to play outdoors.

Including Families

- Host a monthly (or weekly!) hike with families. Hike near your program site or seek out local parks and trails and explore the area together.
- Assign "homework" to families to go outdoors and look for seasonal items or make up an outdoor treasure hunt for families to complete.
- Encourage families to get together for picnics, outdoor play, and adventuring to build community and boost everyone's confidence in outdoor play.

More Information

Start with the following websites to learn more about the research that supports the value of outdoor play. Many of them offer research digests or tear sheets to share with families. You will also find resources for how to safely and regularly go outdoors, tips and tricks for teaching outside, and professional development resources. The Alliance for Childhood (www.allianceforchildhood.org) provides resources for educators and families and offers information and research about national policy related to children and childhood. Natural Start Alliance (www.naturalstart.org) offers a blog and a map of nature-based preschools and early childhood programs, and it has a vast collection of resources, including articles, webinars, how-to guides, and more. The Outdoors Alliance for Kids website (https://outdoorsallianceforkids.org) offers ideas for organizing play events and supporting children's right to play.

CHAPTER 7
EARTH-FRIENDLY EDUCATION

Mr. Lopez supports his teachers in their efforts to promote Earth-friendly attitudes, behaviors, and choices in the classroom. He knows that the practices promoted by teachers have a big impact on children's learning. He also encourages his teachers to use an emergent curriculum based on the children's interests and to choose books and materials that help children appreciate the Earth and all who live here. He hosts family nights where parents and other caregivers get together, strengthen the class community, and learn about conservation and the local environment. He knows that sharing plenty of nature-based materials, stories about nature and animals, and time outside, whether in the school garden or elsewhere, provides many provocations for children's learning. As they learn and experience the natural world for themselves, he knows they will be inspired to care for it.

Mr. Lopez has done a lot of research to support his choices inside the classroom and out. You don't need to be an expert to make eco-friendly choices for your program. Even small steps make a big difference as we teach children to care for the environment.

Why Environmental Education Is Necessary

To date, numerous communities are already feeling the effects of climate change. Low-lying coastal areas of the United States are prone to the destructive effects of hurricanes and tropical storms, and other areas of the country are experiencing record heat, droughts, flooding, wildfires, and other extreme weather events as a result of human-induced climate change. Knowing that young children in America are exposed to news media and footage thanks to the ubiquity of screen-based media in their lives, many children participating in early childhood programs have at least a passing awareness of these events.

Furthermore, children, particularly those who identify as Black, Indigenous, or people of color, make up a huge segment of the population that is affected by environmental racism in the form of disproportionate effects of environmental hazards on their communities. Many industries and organizations deliberately locate hazardous waste sites, polluting factories, and other sources of toxins in neighborhoods that are socioeconomically disadvantaged. This means that environmental issues such as air and water pollution (among others) are a primary part of many children's everyday lived experience. Educators need to balance the need to respond to children's many questions and concerns about environmental issues alongside the need for sensitive, developmentally appropriate teaching methods. Rather than avoid the topic of climate change or dismiss children's fears in an attempt to allay them, educators should be prepared to engage children in activities that allow them to do age-appropriate things to help the Earth.

Within the world of early childhood education, there remains a gray area between early childhood environmental education and education for sustainability. Many people think about *environmental education* as education in, for, and about the environment, whereas *education for sustainability* tends to focus more on the agential role of children, identifying complex environmental issues and identifying ways for young children to feel a sense of agency and control. There is often a disconnect between these two approaches chiefly because some educators are reluctant to engage young learners in practices that burden them with guilt about the environment or create feelings of fear, anxiety, or shame (Elliott, Ärlemalm-Hagsér, and Davis 2020; Sobel 1996). Most educators are understandably concerned about frightening children by teaching them about issues like climate change, pollution, biodiversity loss, and other big problems, but these topics are not unfamiliar to children. There are developmentally appropriate ways to approach sustainability education that support children's connection to and engagement with nature while responding to children's anxiety as needed.

Educators concerned with sustainability tend to focus on conservation education, which means teaching children specific behaviors that preserve and care for the Earth. Often this is done by modeling pro-environment actions, such as caring for living things, picking up trash, recycling, and engaging in other practices described in this book. They also focus on education about and for the environment, meaning that they teach not only the "what" but the "why" about the natural world. For example, rather than just teaching children the names of the plants and animals in the schoolyard, a teacher would spend time helping children understand why those species live in that particular location and how they interact with one another. Another important factor in education for sustainability is helping children to see themselves as agents of change. Many of us feel overwhelmed or anxious when it comes to climate and environmental issues, and early childhood educators want to help children work through those feelings. One way to do so is to engage in activities where children feel like they are making a difference. Like adults, when children have opportunities to act alongside others in ways that benefit the Earth, they feel empowered and less overwhelmed. Research suggests that a key factor in helping children feel empowered and capable of making positive change is acting together with others. This is one reason why being a part of an Earth-friendly classroom can have such a positive impact on a child. They know that they are a part of a community of children and adults who care about the Earth and are doing lots of things to take care of it.

Educating for Sustainability

It might be helpful to think about education for sustainability as having four major themes:

1. *Instilling pro-environment attitudes*, which children develop through spending time caring for nature and being aware of the positive effects of their choices.
2. *Providing education about and for the environment*, which often happens either through outdoor play and learning or through activities, themes, or projects focused on the environment. Targeted information promotes awareness, enjoyment, wonder, and gratitude for the natural world.
3. *Implementing and modeling sustainable practices*, such as recycling, conserving water, and healthy eating and physical activity habits, which help to normalize environmental behaviors.
4. *Foregrounding children as change agents*, involving them in directly caring for the natural world as they take action and demonstrate respect for it.

Even very young children love to help solve problems, so it's important to include shared ideas for saving the environment. On the other hand, it's not fair to burden young children with long lists of environmental choices they can't control, such as "change the lightbulbs," "don't use pesticides on the lawn," and other decisions best left to adults. Child-friendly contributions should be joyful, collaborative work. Think about the environmentally necessary work of your center: Can children help take care of the site? Can they sweep the sidewalks and weed the gardens? Can they refill bird feeders and water plants? Engage them in working together to care for the classroom plants and pets. Have small teams "adopt" the compost or the garden for periods of time and help them be responsible for maintaining the areas. Whole-class activities such as caring for the schoolyard by picking up trash show children that "we're all in this together" and can be done as an act of love and gratitude for your site and the plants and animals that surround it.

Direct and Indirect Nature Experiences

Activities that engage children in the natural world, whether directly or indirectly, can be subtle yet effective ways of teaching sustainability. Here are some examples:

Direct Experience	Indirect Experience
• hiking in nature • going on a picnic • working in the garden • filling a bird feeder • watering plants and trees • raking leaves • sweeping sidewalks, shoveling/playing in snow • reading or having snack outside • playing outdoors away from the playground	• playing with natural loose parts • drawing or painting pictures about nature • dictating stories about nature • reading nature stories • looking at pictures and posters of nature scenes • looking out windows • imagining being outside • engaging in guided meditation about being outdoors

Use emergent curriculum approaches to respond to children's curiosity about the natural world. Practice listening to the class's interests and using those topics to spark investigations or inspire long-term dramatic play and class projects. Venturing outdoors often will give you and the children lots to be inquisitive about, which will lead to questions that may set the whole class down a path of discovery.

Books that have nature and animals as themes are particularly appealing to young children for a number of reasons, including children's innate curiosity about animals and the places where they live, their developing sense of self-to-other relations, and their growing understanding of the world around them. Teachers often turn to environmental literature with specific developmental outcomes in mind: to support children's literacy development, to help children understand and respond to environmental problems, and to support children's sense of agency. We all know how important feelings are, so make story time positive and curiosity-rich to help pique children's interest in the natural world.

Health and Safety Considerations

- When you talk about the environment, be positive and upbeat, but don't be dismissive of children's fear or anxiety. As much as we want to protect children from the scary realities of climate change, many of them know and feel more than we realize. Remember the techniques discussed here for helping children develop a sense of agency, and remind them that they are safe and they can do lots of things to help the Earth.
- If you are especially concerned about a child's anxiety, treat the situation as you would any other concerning one: work with the child's caregivers, a social worker, and other professionals trained in trauma-informed approaches.

Practical Applications

- Incorporate nature-themed storybooks into your literature collection. (Take a look at "our big list of beloved books" at Natural Start: https://naturalstart.org/feature-stories/beloved-books.)
- Help the children document and communicate their feelings about nature: take digital photos, sketch, or describe the places they like to visit around the schoolyard and create a classroom book. Creating names for children's special places outdoors, such as The Big Shady Tree, The Rolling-Down Hill, and so on, can have positive impacts. Take good care of these places by visiting them often.
- Normalize pro-environmental behaviors in your program: conserving water, reducing waste, recycling paper, and so on. Modeling shows children just how easy these behaviors can be. Children feel good when they work together to solve problems.

More Information

The Natural Start Alliance, a program of the North American Association for Environmental Education (www.naaee.org/naturalstart) is a network of nature-based early learning programs across the United States and beyond, offering program support, resources, and ideas for administrators and teachers, webinars, recommendations, and learning materials for everyone concerned with going green in early childhood: parents, teachers, and administrators alike.

The Children and Nature Network (www.childrenandnature.org) has an online research library that offers collections of peer-reviewed research on the benefits of teaching outdoors as well as research briefs—short documents outlining the many benefits of outdoor learning for young children.

CHAPTER 8 CONCLUSION

There is no denying that climate change is real, but we can take steps to become resilient and sustainable. The only way toward a more just and sustainable future is to take action. Some ideas in this book ask you to make sacrifices of convenience or time. Those sacrifices may feel challenging, inspiring, or daunting. Some may feel more manageable. The important thing is to do something. Challenge yourself and your colleagues to take meaningful action to shift your program in an Earth-friendly direction.

As this book has shown, you can make a wide variety of choices in becoming more ecological and sustainable. Take the first step with some reflection as a team on how your program defines "Earth friendly" and what steps are reasonable to take. Talk with your colleagues or reflect on your own. Do you want to overhaul your food program? Start a school garden? Great. Maybe doing research on product life cycles and seeking out the lowest-impact choices is more your style. That's important work too. Remember that moving toward sustainability is a journey for everyone, and every journey begins with a simple step.

APPENDIX A
PURCHASING CRITERIA

Use this set of criteria as a tool to help you make decisions about new materials before you purchase them. Based on the product life cycle graphic on page 20, this checklist takes into account a number of factors to help you evaluate where a material fits on your program's Earth-friendly spectrum. Before going through these steps, talk with your staff to determine whether the materials are needs or wants.

1. Is there any way this need could be met with existing resources that I can repurpose or reuse?

2. If I must get this item new, what is it made of?

    - Nonrenewable resources (such as plastic made from oil)?
    - Sustainably produced materials?
    - Some recycled materials?
    - Completely recycled materials?

3. Does this product contain minerals that were mined from the Earth?

4. Where was it produced?

5. Was the process by which it arrived here one that required lots of fossil fuel?

6. Is this a product that can be used over and over, or is it single use?

7. Does the label for this product contain any signal words (see list on p. 14)?

8. What will happen to this product at the end of its life? Can it be recycled, or is it biodegradable?

APPENDIX B SAMPLE ENVIRONMENTALLY PREFERABLE PURCHASING PLAN

Sunny Days Preschool

Purpose: We recognize that we all have a role to play in working to create a safe and healthy future for the children in our care. For that reason, we've established a set of parameters to help guide our purchasing decisions.

These parameters will help us

conserve natural resources through reduction of purchasing, reusing where possible, and recycling;

reduce the consumption of water and electricity;

reduce food waste;

support locally grown or produced foods and materials, thereby supporting our community;

reduce the consumption of single-use plastics and other materials destined for landfills;

model environmentally friendly business practices; and

educate our community, including our staff, vendors, and the children and families we serve.

Important Definitions

environmentally preferable product: a product that has a lesser or reduced negative effect on human health and the environment when compared to other items that serve the same purpose

life cycle: the resources associated with a product throughout its entire lifetime, including extraction of natural resources, transportation, manufacturing, packaging, use, and disposal

recyclable: a product that, after its intended use, can be broken down and used as a raw material in the manufacture of another product

reusable: a product that can be used several times for an intended use or used for a different purpose before being discarded

APPENDIX C
EARTH-FRIENDLY PROGRAM CHECKLIST

This checklist is designed to help you identify areas where you're already making Earth-friendly choices and where you have room to grow. You can use these guidelines as inspiration for goal setting. Choose one area of focus, or use them all. It will give you an opportunity to reflect on your program's current practices related to each of the sections in this book.

After each statement below, fill in the circle from "not really" to "completely" that best shows where your program is currently on the spectrum.

General

As a team, we are aware of local environmental issues and the communities most affected by climate change.

not really O O O O O O O completely

As a team, we are committed to honoring every child's right to clean air and water, space to play, and freedom to grow up healthy.

not really O O O O O O O completely

Indoor Environment

We prohibit vehicle idling outside our doors during pickup, drop-off, or any other times.

not really O O O O O O O completely

Smoking is prohibited on the premises.

not really O O O O O O O completely

The chemicals we use for cleaning, disinfecting, and sanitizing are certified as Earth friendly by a third-party organization such as the Environmental Working Group.

not really O O O O O O O completely

We avoid aerosol products (such as cleaners, deodorizers, or sunscreen).

not really O O O O O O O completely

We model Earth-friendly self-care, such as frequent handwashing, healthy eating, and drinking water.

not really O O O O O O O completely

We regularly engage in physical activities with children to promote physical fitness.

not really O O O O O O O completely

We are aware of the rules and regulations as they apply to cleaning and sanitizing products and practices.

not really O O O O O O O completely

We will replace one or more of our cleaning supplies with less-toxic alternatives.

not really O O O O O O O completely

We rinse tables and supplies after using cleaning products on them.

not really O O O O O O O completely

Reduce, Reuse, Recycle

We reuse art supplies whenever possible.

not really O O O O O O O completely

We have a plan to reduce our consumption of single-use plastics.

not really O O O O O O O completely

As a team, we are aware of the life cycle of many of our most common supplies and equipment.

not really O O O O O O O completely

We know which plastics and other materials are recyclable in our area.

not really O O O O O O O completely

When we do purchase new products such as supplies or equipment, we recycle all packaging.

not really O O O O O O O completely

We carefully consider a product's life cycle and buy new products only when absolutely necessary.

not really O O O O O O O completely

We host regular "swap meets" where families can trade outgrown clothing or toys with one another.

not really O O O O O O O completely

Learning Spaces

We have access to a safe natural area in which to play.

not really O O O O O O O completely

The children have opportunities to care for the Earth through gardening, feeding birds and other wildlife, and planting trees, shrubs, or flowers.

not really O O O O O O O completely

In the classroom, we have a collection of loose parts collected from the outdoors so that children have contact with nature even while indoors.

not really O O O O O O O completely

We have houseplants in the classroom.

not really O O O O O O O completely

We take "indoor" activities like story time, snacktime, or circle time outdoors.

not really O O O O O O O completely

We regularly go outdoors to play for at least five hours per week.

not really O O O O O O O completely

We walk around the local area so that the children learn more about where they live and learn.

not really O O O O O O O completely

Earth-Friendly Indoor Play

We conserve water and learn about water mostly through outdoor play rather than using a traditional water table, which wastes water.

not really O O O O O O O completely

We evaluate our art materials to ensure they are nontoxic and Earth friendly.

not really O O O O O O O completely

We avoid glitter.

not really O O O O O O O completely

We create musical instruments from reused materials, or we have musical instruments made from natural materials like shells, seedpods, or reeds.

not really O O O O O O O completely

If we have plastic blocks, they are frequently sanitized and of a high enough quality to be reused for many years to come.

not really O O O O O O O completely

We have toys made from wood or other natural materials.

not really O O O O O O O completely

Earth-Friendly Curricula

We are aware of the environmental issues local to our program, and we address them in our curricula.

not really O O O O O O O completely

We encourage children to talk about their feelings about climate change and the environment.

not really O O O O O O O completely

We seek out books and other materials that celebrate the natural world rather than presenting it as a place of doom and gloom.

not really O O O O O O O completely

We offer frequent opportunities for children to engage in the natural world both directly and indirectly.

not really O O O O O O O completely

We are consistent with our sustainability practices, and we model them for children to see.

not really O O O O O O O completely

We support children's agency and their desire to take action on behalf of the environment.

not really O O O O O O O completely

We support children's feelings and questions about the natural world.

not really O O O O O O O completely

Earth-Friendly Snack and Mealtimes

We purchase organic, locally grown fruits, vegetables, and legumes for snack and meals.

not really O O O O O O O completely

We avoid serving meat due to the impact of animal agriculture.

not really O O O O O O O completely

We have a school garden or containers in which we grow vegetables or herbs.

not really O O O O O O O completely

We recycle food packaging and seek to avoid single-use containers.

not really O O O O O O O completely

We are aware of the rules related to serving and storing food at our center, and we have determined ways to reduce or eliminate plastics and chemically processed food.

not really O O O O O O O completely

What other steps are you taking?

What are a few small changes you could make now to be more Earth friendly?

What are a few small changes you could make in six months? One year?

Who can you engage for support as you shift toward a more Earth-friendly program?

- Custodial staff?
- Caregivers?
- Community members?

- State and local resources (licensing, agricultural agency or extension service, and so forth)
- Colleagues?
- Administrators?

How will you engage these folks?

REFERENCES

Chawla, Louise, and Victoria Derr. 2012. "The Development of Conservation Behaviors in Childhood and Youth." In Susan D. Clayton, ed. *The Oxford Handbook of Environmental and Conservation Psychology.* New York: Oxford University Press.

Elliott, Sue, Eva Ärlemalm-Hagsér, and Julie Davis, eds. 2020. *Researching Early Childhood Education for Sustainability: Challenging Assumptions and Orthodoxies.* New York: Routledge.

EPA (United States Environmental Protection Agency). 2009. "Persistent Organic Pollutants: A Global Issue, A Global Response." www.epa.gov/international-cooperation/persistent-organic-pollutants-global-issue-global-response.

——. 2021a. "Containers and Packaging: Product-Specific Data." Accessed December 27, 2021. www.epa.gov/facts-and-figures-about-materials-waste-and-recycling/containers-and-packaging-product-specific-data.

——. 2021b. "Learn about Environmental Justice." Accessed December 27, 2021. www.epa.gov/environmentaljustice/learn-about-environmental-justice.

——. 2021c. "Tool for Reduction and Assessment of Chemicals and Other Environmental Impacts (TRACI)." Accessed December 27, 2021. www.epa.gov/chemical-research/tool-reduction-and-assessment-chemicals-and-other-environmental-impacts-traci.

EWG (Environmental Working Group). 2021. "EWG's 2021 Shopper's Guide to Pesticides in Produce™." www.ewg.org/foodnews/summary.php.

FAO (Food and Agriculture Organization of the United Nations). 2013. "Tackling Climate Change through Livestock." www.fao.org/ag/againfo/resources/en/publications/tackling_climate_change/index.htm.

Fuhrman, Joel. 2018. "The Hidden Dangers of Fast and Processed Food." *American Journal of Lifestyle Medicine* 12 (5): 375–81. http://doi.org/10.1177/1559827618766483.

Hever, Julieanna. 2016. "Plant-Based Diets: A Physician's Guide." *The Permanente Journal* 20 (3): 15–082. https://doi.org/10.7812/TPP/15-082.

Kuo, Ming, Michael Barnes, and Catherine Jordan. 2019. "Do Experiences with Nature Promote Learning? Converging Evidence of a Cause-and-Effect Relationship." *Frontiers in Psychology* 10 (February 19). https://doi.org/10.3389/fpsyg.2019.00305.

Lanphear, Bruce P. 2015. "The Impact of Toxins on the Developing Brain." *Annual Review of Public Health* 36 (March): 211–30. https://doi.org/10.1146/annurev-publhealth-031912-114413.

Lappé, Anna. 2010. *Diet for a Hot Planet: The Climate Crisis at the End of Your Fork and What You Can Do about It*. New York: Bloomsbury.

Moore, Robin C. 2014. *Nature Play and Learning Places: Creating and Managing Places Where Children Engage with Nature*. Raleigh, NC: Natural Learning Initiative and Reston, VA: National Wildlife Foundation. http://outdoorplaybook.ca/wp-content/uploads/2015/09/Nature-Play-Learning-Places_v1.5_Jan16.pdf.

North American Association for Environmental Education (NAAEE). 2010. *Early Childhood Environmental Education Programs: Guidelines for Excellence*. Washington, DC: NAAEE.

Panoff, Lauren. 2021. "I'm Raising My Kids on a Plant-Based Diet for Their Future." Healthline Perspective. Last medically reviewed June 8, 2021. www.healthline.com/nutrition/raising-kids-vegetarian#environmental-benefits.

Quinn, Cristina L., and Frank Wania. 2012. "Understanding Differences in the Body Burden-Age Relationships of Bioaccumulating Contaminants Based on Population Cross Sections versus Individuals." *Environmental Health Perspectives* 120 (4): 554–59. doi:10.1289/ehp.1104236.

Selly, Patty Born. 2012. *Early Childhood Activities for a Greener Earth*. St. Paul, MN: Redleaf Press.

Sobel, David. 1996. *Beyond Ecophobia: Reclaiming the Heart in Nature Education*. Great Barrington, MA: Orion Society.

Steinfeld, Henning, Pierre Gerber, Tom Wassenaar, Vincent Castel, Mauricio Rosales, and Cees de Haan. 2006. *Livestock's Long Shadow: Environmental Issues and Options*. Rome, Italy: Food and Agriculture Organization of the United Nations. www.fao.org/3/a0701e/a0701e.pdf.

Sullivan, Laura. 2020. "How Big Oil Misled the Public into Believing Plastic Would Be Recycled." *NPR Morning Edition*, September 11, 2020. www.npr.org/2020/09/11/897692090/how-big-oil-misled-the-public-into-believing-plastic-would-be-recycled.

United Nations Children's Fund. 2017. "17 Million Babies under the Age of 1 Breathe Toxic Air, Majority Live in South Asia." www.unicef.org/press-releases/babies-breathe-toxic-air-south-asia.

——. 2021. "Preventing a Lost Decade." www.unicef.org/media/112891/file/UNICEF%2075%20report.pdf.

United Nations Environment Programme. 2018. "Young and Old, Air Pollution Affects the Most Vulnerable." Blog post, October 16, 2018. www.unep.org/news-and-stories/blogpost/young-and-old-air-pollution-affects-most-vulnerable.

University of Arizona. 2019. "Buying Less Is Better Than Buying 'Green'—for the Planet and Your Happiness." *ScienceDaily*, October 8. www.sciencedaily.com/releases/2019/10/191008155716.htm.

USDA and USHHS (United States Department of Agriculture and U.S. Department of Health and Human Services). 2020. *Dietary Guidelines for Americans, 2020–2025*. 9th ed. December 2020. www.DietaryGuidelines.gov.

Weatherly, Lisa M., and Julie A. Gosse. 2017. "Triclosan Exposure, Transformation, and Human Health Effects." *Journal of Toxicology and Environmental Health Part B, Critical Reviews* 20 (8): 447–69. doi:10.1080/10937404.2017.1399306.